La Posada Hotel Fire
McAllen, Texas

Investigated by: Jeffrey M. Shapiro, P.E.

This is Report 001 of the Major Fires Investigation Project conducted by TriData Corporation under contract EMW-88-C-2277 to the United States Fire Administration, Federal Emergency Management Agency.

Homeland Security

Department of Homeland Security
United States Fire Administration
National Fire Data Center

U.S. Fire Administration Fire Investigations Program

The U.S. Fire Administration develops reports on selected major fires throughout the country. The fires usually involve multiple deaths or a large loss of property. But the primary criterion for deciding to do a report is whether it will result in significant "lessons learned." In some cases these lessons bring to light new knowledge about fire--the effect of building construction or contents, human behavior in fire, etc. In other cases, the lessons are not new but are serious enough to highlight once again, with yet another fire tragedy report. In some cases, special reports are developed to discuss events, drills, or new technologies which are of interest to the fire service.

The reports are sent to fire magazines and are distributed at National and Regional fire meetings. The International Association of Fire Chiefs assists the USFA in disseminating the findings throughout the fire service. On a continuing basis the reports are available on request from the USFA; announcements of their availability are published widely in fire journals and newsletters.

This body of work provides detailed information on the nature of the fire problem for policymakers who must decide on allocations of resources between fire and other pressing problems, and within the fire service to improve codes and code enforcement, training, public fire education, building technology, and other related areas.

The Fire Administration, which has no regulatory authority, sends an experienced fire investigator into a community after a major incident only after having conferred with the local fire authorities to insure that the assistance and presence of the USFA would be supportive and would in no way interfere with any review of the incident they are themselves conducting. The intent is not to arrive during the event or even immediately after, but rather after the dust settles, so that a complete and objective review of all the important aspects of the incident can be made. Local authorities review the USFA's report while it is in draft. The USFA investigator or team is available to local authorities should they wish to request technical assistance for their own investigation.

For additional copies of this report write to the U.S. Fire Administration, 16825 South Seton Avenue, Emmitsburg, Maryland 21727. The report is available on the Administration's Web site at http://www.usfa.dhs.gov/

U.S. Fire Administration

Mission Statement

As an entity of the Department of Homeland Security, the mission of the USFA is to reduce life and economic losses due to fire and related emergencies, through leadership, advocacy, coordination, and support. We serve the Nation independently, in coordination with other Federal agencies, and in partnership with fire protection and emergency service communities. With a commitment to excellence, we provide public education, training, technology, and data initiatives.

TABLE OF CONTENTS

LA POSADA HOTEL FIRE
McAllen, Texas – February 25, 1987

Investigated by: Jeffrey M. Shapiro, P.E.

Local Contacts: Chief Butch Derr
F.M. Honore Castro
Lt. Richard Loza
City of McAllen Fire Department
101 South Bicentennial
McAllen, Texas 78501
(512) 631-3301

OVERVIEW

On February 25, 1987, the successful operation of a smoke detector provided the early warning that mitigated a potential multiple death fire in the La Posada Hotel in McAllen, Texas. Though the fire caused approximately 150,000 dollars in damages, 155-160 occupants were evacuated with only one serious injury. The fire was caused by smoking in bed and sent heavy smoke throughout most of the hotel. Key negative factors in the fire included the lack of self-closing doors between guest rooms and corridors, open stairwells, and the lack of sprinklers and fire alarms systems. Key positive factors included smoke detectors in each room, light fire loading, fire resistive construction, and an unusually high proportion of occupants familiar with the hotel.

BACKGROUND

The La Posada Hotel is a designated historic structure registered with the State of Texas. The original structure was built in 1918. In 1973, the building was struck by lightning which caused a major fire that destroyed the original structure with the exception of some exterior walls. The rebuilt building is three stories tall with 164 guest rooms plus meeting facilities. (See Figure 1.)

Codes

McAllen currently uses both the Southern Standard Building Code and the full set of National Fire Codes. Based upon the timeframe for construction provided by fire department sources, the building probably was constructed to the 1973 Southern Standard Building Code. Additionally, National Fire Protection Association (NFPA) 101, Life Safety Code, may have been applied. Though NFPA 101 contains provisions for safety in existing buildings, no retroactive enforcement program had been established.

Construction

The building was completely reconstructed using non-combustible, fire resistive construction. The structural frame consists of concrete double-tee floor assemblies and masonry block walls. The overall fire protection classification appears to be Type II, fire resistive. All guest rooms are separated from other guest rooms, corridors, and public areas by at least one-hour fire resistive construction with the exception of corridor doors.

Corridor doors to guest rooms are panel-style wood doors, non-rated, and without closing devices. These doors are hung in solid wood frames.

Exits

Exiting from the guest rooms was through interior corridors or through an operable window in each room. (See Figures 2 and 3.) The building has four interior exit stairways, all of which are unenclosed or inadequately enclosed, based upon the criteria for new and existing hotels contained in NFPA 101, Life Safety Code. (See Figures 4 and 5.)

Interior Finishes

All interior wall surfaces are painted. Floors in guest rooms and guest room access corridors are carpeted. (See Figure 6.)

Fire Protection Systems and Equipment

The only fire protection equipment in the hotel consisted of 2 ½ gallon pressurized water fire extinguishers, 1 ½ inch occupant use hose cabinets supplied by domestic water only, and battery-powered single station smoke detectors in guest rooms and storage areas. The smoke detectors are General Electric SMK-6/M1. They were located above the ceiling in the return air plenum for the room. (See Figure 7.) This location is not the recommended location according to NFPA 72E, standard on automatic fire detectors.

Fire Department

McAllen Fire Department operates six fire stations will 111 personnel. The department operates six engines, one truck, one rescue, and a crash-fire-rescue vehicle.

THE FIRE

The fire originated in Room 123 in the north wing of the hotel. That wing contained 22 guest rooms and two storage rooms on the first floor.

According to the room occupant's statement to fire investigators, he returned to the hotel at approximately 11:00 p.m., under the influence of alcohol and lied down on the bed to watch television while smoking. He fell asleep and the cigarette ignited a small area of bedding. The smoke detector alarmed at approximately 1:00 a.m.

At the time of the fire only two hotel staff members, a night auditor and a janitor, were on duty. Having heard the smoke alarm from the front desk, which was located approximately 50 feet down the corridor from Room 123, the two staff members went into the corridor and determined that the alarm was in Room 123. They then went back to the desk and called that room. The occupant said everything was OK, and no further action was taken.

Approximately 15 minutes later, the occupant called the front desk to advise that there had been a fire and that there was some damage. He requested that a staff member come assess the damage so that he could make restitution. The night auditor advised that the hotel would handle the matter in the morning when the occupant checked out.

The occupant then removed the mattress from the box spring and placed the mattress vertically against the wall adjacent to the bed, leaning it against the window curtains. He then opened the window to remove the smoke odor unknowingly providing a fresh air source to the mattress which was still smoldering. The occupant then went back to sleep on the box spring.

Sometime after 4 a.m., the occupant was again awakened by the smoke detector and discovered flames involving the mattress and spreading to the curtains. He went outside the room leaving the door open behind him, went to the front desk, and advised the auditor to call the fire department. He went back to the room with the janitor and a 2 ½ gallon pressurized water fire extinguisher. At this point, flames were rolling off the ceiling. The janitor first used the extinguisher with no success. The occupant then took the extinguisher and crawled into the room to attempt extinguishment. The extinguisher expired, and the occupant backed out leaving the door open behind him. (See Figures 8 and 9.)

The fire began to extend out the doorway into the corridor. (See Figure 10.) The janitor went to the second and third floors and began awakening guests. Many guests came into the corridors to check the commotion. Fire was continuing to extend towards the front desk on the first floor level, and smoke was spreading throughout the corridors in the north, east, and west wings of the hotel on all three floors. On upper floors, smoke was particularly heavy in the north and east wings having traveled primarily up the open "grand" stairway in the lobby. Glass stairway doors separating the other stairways from the first floor corridor remained in-place for between five and ten minutes before failing. The hotel's assistant management was on the premises and was notified of the fire immediately after the fire department was called. She left her room and attempted egress through an interior stairway. Upon opening the glass door to the first floor corridor, she was exposed to heavy smoke. Smoke spread was so rapid that by the time she was able to return to the second floor, it too was smoke-filled, and she was forced to break a window with a chair to escape to an outside balcony.

The fire department received its first notification of the fire by telephone from the night auditor at 4:13 a.m. The first dispatch was made sending a first alarm assignment at 4:13 with the first units arriving on the scene at 4:14. Initial dispatch consisted of two engines, one truck, one rescue units, and one deputy chief, a total of 11 personnel. The first unit arriving noted fire in the lobby, corridor, and the room of origin. As yet, many occupants were still unaware of the fire. Others were evacuating through smoky corridors or awaiting rescue at windows. Given the number of occupants at windows requiring rescue, the initial fire department operations were geared to ground ladder rescues. Actual firefighting was limited to one handline placed into service at the room of origin through the exterior window by a single firefighter. Later, a second handline was placed in service from the interior. Though the fire was extinguished relatively quickly, the size of the structure, the number of occupants, and the extent of smoke spread necessitated fire department operations escalating to a third-alarm.

Of the 164 guest rooms available for rental at the time of the fire, 125 were occupied. The estimated total number of occupants in the building at the time of the fire was 155-160. According to fire department reports, approximately 15 people were rescued by ground and aerial ladders, and the remainder escaped on their own or were escorted out through smoky halls and interior stairway exits by fire department personnel.

The only serious injury reported was that of a woman who was sitting on a window ledge on the third floor calling for help when the fire department arrived, and either jumped or fell. Fire damage was limited to the room of origin, approximately fifty feet of corridor space in either direction, and a small area at the front desk. Smoke damage was considerable throughout the north, east, and west wings.

ANALYSIS OF SIGNIFICANT FACTORS

Building Construction and Contents

The combination of a light fire loading and non-combustible, fire resistive building construction helped limit the fire damage to the room of origin and the corridor in the immediate vicinity. Combustible contents in the room of origin were limited to a wood desk, two wood chairs with small seat and/or back pads, a wood dresser, a television with stand, drapes, and a bed. (See Figure 3.) The mattress consisted of 30 percent urethane foam, 40 percent blended cotton felt, and 30 percent sisal. (See Figure 11.) It was apparently purchased only three years ago which would indicate compliance with Federal smolder resistance standards. However, the testing for compliance with Federal standards does not contemplate a vertical mattress placed against combustible curtains next to an open window. The bedding, furniture upholstery, and window curtains were unlabeled and probably ignited easily. The only combustible materials in the corridor were the carpeting and decorative wood beams on the non-combustible ceiling. The lack of rated self-closing doors to guest rooms allowed smoke and fire to escape into the corridor from the room of origin, and the lack of stairway enclosures allowed smoke to spread to the upper floors.

Fire Protection Equipment

A key factor in the successful evacuation of occupants was the early warning provided by the smoke detector in the room of origin. Had the fire been full extinguished when first detected, a major fire incident would not have occurred; however, the early detection of the rekindle was ultimately responsible for the escape of the room occupant, the early notification of the fire department, and the evacuation of guests. Texas passed a law in 1983 which required all hotels, motels, and apartments to provide smoke detectors to protect sleeping areas.

The fire department did experience some trouble evacuating the building due to the lack of a fire alarm system. Many guests were unaware that there was a fire until fire department personnel knocked on the doors to evacuate them. Some actually refused to leave until ordered to do so by police.

The lack of sprinklers and fire alarm systems placed additional demands on the fire department to rescue occupants. The fire department was able to gain control of the situation due to prompt notification and fast response times. Had sprinklers been present in the hotel's guest rooms, it appears that the fire might well have been controlled by one sprinkler head.

Human Behavior

Unlike the transient nature of occupants in a typical hotel, the La Posada was occupied almost entirely by guests who were somewhat familiar with the hotel. Eighty-five percent of the guests were with the Federal Deposit Insurance Corporation, and had been in the hotel for one and a half weeks at the time of the fire. Of the other 15 percent, many were seasonal guests who had been residing in the hotel for an extended period. Occupant familiarity is thought to have contributed to the large number of guests who were able to evacuate on their own, and probably eased the anxiety of those evacuated by the fire department in smoky conditions.

LESSONS LEARNED

The La Posada fire demonstrated once again the value of smoke detectors in providing early, life-saving warning of fire. A working detector and the quick response of the fire department, coupled with their tactical decision to focus staffing on safely evacuating the occupants, were the main, positive lessons drawn from this fire. These points are discussed below:

Smoke Detectors are Critical for Life Safety

Not once, but twice, the smoke detector in Room 123 alerted the occupant to a developing emergency. Unwise actions by the occupant and lack of a suitable response by the hotel's night employees after the first warning led to a rekindling of the mattress and the need for the second sounding of the detector. Given some of the other conditions present, e.g., lack of an alarm system, unenclosed/inadequately enclosed stairways, and the late hour the fire occurred (guests were sleeping and unaware), the losses in all likelihood would have far exceeded the one injury and 150,000 dollars damage had it not been for the presence of a working detector.

Ironically, the detector was not properly located. It had been installed (as had the other guest room detectors) in the return air plenum, rather than on the ceiling of the room. Moreover, the detector was a battery-operated type, not hard-wired to the electrical system as is preferred for this type of occupancy.

Fire Officer Decision to Make Evacuation the Priority Lessened Risk of Greater Losses

The decision to assign ten or eleven of the personnel on the first-arriving units to alert and evacuate occupants probably averted additional civilian casualties. Since there was no alarm system in the building, firefighters had to go door-to-door, a time consuming job, before sleeping occupants could be awakened and helped to safety. The rapid spread of smoke escalated the danger, and more casualties might have occurred had the occupants not been warned to get out. As more units arrived, another 25 firefighters became available to rescue and evacuate people. Eight firefighters from one of the three mutual aid companies that responded also were assigned rescue duties.

In retrospect, it would have been a good idea for firefighters to have gotten master keys at the outset so they could have alerted guests more quickly and avoided kicking in the guest room doors. (Keys were obtained after awhile.) Nevertheless, firefighters otherwise showed awareness of proper procedures, e.g., they propped open doors of rooms, already evacuated, established a check-point, oversaw the return to rooms and occupants' removal of their belongings, etc.

What lessons can be learned about areas for improvement? Most of the problems centered around certain building features and human behaviors that contributed to the start and spread of the fire. There are a number of key factors under both headings that should be examined.

Improper/Inadequate Building Features Helped Spread Fire

Once again, we see the impact that building features can have either in containing or aiding the spread of fire and smoke. Among the more serious problems were:

- Guest room doors did not have automatic closures and were constructed of wood. The door to the room of origin was left open.

- Stairways leading to the second and third levels were not enclosed or were inadequately enclosed. Smoke spread through them.

- There was no central fire alarm system. Occupants had to be alerted by firefighters going door-to-door.

- There were no automatic sprinklers. Fortunately, fire-resistive construction and low fire load in the corridor helped check the spread of flames.

- A few rooms on the first floor had burglar bars installed over the exterior windows. However, the fire department stated that the bars were easily removed and did not hamper rescue. Obviously, burglar bars are an impediment to emergency egress and, when installed, should have a clearly identified emergency egress release which occupants can use in an emergency. Such release devices are required by current codes.

Lack of Awareness of Fire Safety Rules Caused the Fire and Prevented Early Extinguishment

Human carelessness caused the La Posada fire. The occupant of Room 123 broke one of the cardinal rules in fire safety – not to smoke in bed. Other mistakes were also made:

- The occupant opened the window near where he placed the smoldering mattress.

- Night shift employees, after having the first detector warning, accepted the guest's reassurance that nothing was wrong. Even when the occupant called 15 minutes later to report minor fire damage, they failed to check out the situation further.

- The occupant left his room door open when he went to advise front desk personnel of the rekindled fire. After he and the janitor attempted to extinguish the fire, they left the room with the door hanging open.

These human behavior factors were, in large measure, responsible for the fire. Fire safe behaviors would have prevented the fire in the first place.

CONCLUSION

The combination of early warning, limited combustible contents, occupant familiarity, and fast fire department action in evacuation were the prime factors which contributed to the successful out-come of the fire. The addition of a sprinkler system, a fire alarm system, rated self-closing doors, and enclosed stairwells could have helped to reduce losses and the risk to the occupants. All of these recommendations are included in the requirements or request of the McAllen Fire Department for reconstruction of the hotel.

Further, losses and risk may have been reduced if a fire safety plan had been developed. A proper hotel fire safety plan, if provided, would likely have included training which instructed the staff to have called the fire department immediately upon an indication of a potential fire. In addition, such a plan would likely have contained provisions to prohibit a guest from returning to a fire area to fight a fire and provisions for evacuation of guests in a fire emergency.

SUPPLEMENTAL INFORMATION

Hotel Owner:

Huntington Hotel Corporation
100 Crescent Court, 17th Floor
Dallas, Texas 75201
(214) 954-1700
Contact: Scott Lynch (should have information on interior finishes, furnishings, and appointments used in hotel.)

On-Site Hotel Representative:

Martha Estes
Assistant Manager
La Posada Hotel
100 North Main Street
McAllen, Texas 78501
(512) 686-5411

Construction plans available through: Turner and Morales Architect, McAllen, Texas

Figure 1. La Posada Hotel (view of the east end main entrance).

Figure 2. Typical interior corridor.

Figure 3. Guest room identical to room of original. Note exterior window and contents.

Figure 4. Main entry foyer. The front door is to the right. The wall in the rear of the picture was erected after the fire to separate the damaged area. During the fire, smoke traveled past the front desk (visible in the lower portion of the wall behind the telephone operator) and up the "grand" stairway to the upper floor corridors.

Figure 5. Typical open stairway (view
from a second floor corridor).

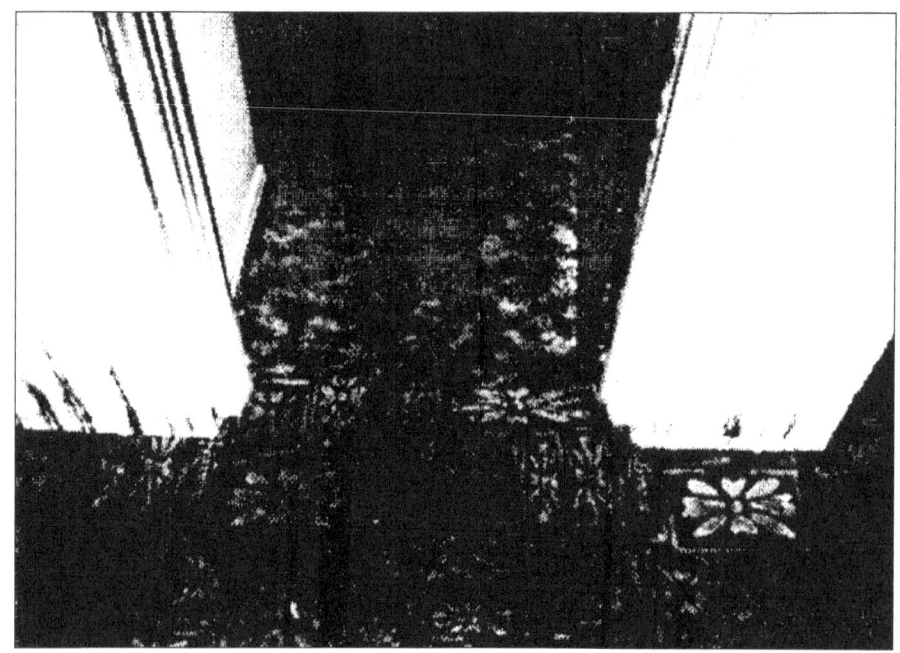

Figure 6. Carpeting in interior corridor and guest room. Also note
wood panel door without door closer.

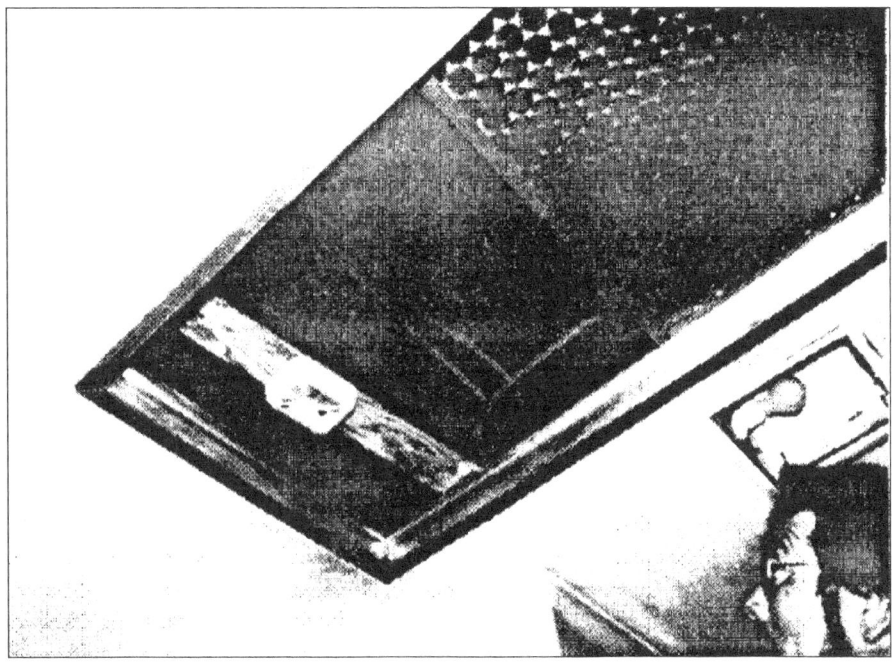

Figure 7. Battery powered smoke detector in adjacent room. All detectors were located in return air plenum above filters and grilles in ceiling.

Figure 8. Room of origin. Mattress was standing up against wall with window.

Figure 9. Room of origin.

Figure 10. View from the front desk into the corridor towards the
room of origin. Note the descending smoke mark level on wall
towards the area of origin.

Figure 11. Mattress identification tag.

Appendix A

1. Texas Fire Incident Reporting System Incident Report

2. McAllen Fire Department Dispatcher's Incident Report

3. Floor Plan Indicating First Floor Layout and Damage

4. Site Plan Indicating Fire Department Deployment

5. Narrative Reports of Fire Department Personnel

6. Recommendations and Requirements for Reconstruction

TEXAS FIRE INCIDENT REPORTING SYSTEM
Incident Report

MCALLEN   Fire Department

902 F 6/77
Layout 2

TEXFIRS INCIDENT REPORT

1 ☐ Delete
2 ☐ Change

Fill in This Report In Your Own Words

A FDID   KN608   Incident No. 0 0 0 2 5 7   Exp No. 0 0   Mo. 0 2   Day 2 5   Year 8 7   Day of the Week WEDNESDAY   Alarm Time 0 4 1 3   Time in Service 1 1 4 7

B CORRECT ADDRESS   No. 1 0 0   Dir.   Name NORTH MAIN   Type   Zip Code 7 8 5 0 1   Census Tract 0 2 7 7 0 0

C Occupant Name -- LA POSADA -- HOTEL   Telephone 686 5411   Room or Apt 123

D Owner Name Womack / Gilman Interest   Address 1001 TEXAS SUITE 600 Houston, Texas   Telephone 713-224-7166

E Method ~ Alarm From Public 512-687-1111   1 1   Type of Situation Found STRUCTURE FIRE   1 1 1

F Type of Action Taken EXTINGUISHMENT   1 1   Co. Inspection District 0 0 1   Shift C   No Alarms 3   Mutual Aid 1 ☐ Rec'd 2 ☐ Given

G No Fire Service Personnel Used at Scene 0 4 7   No Engines Used at Scene 0 0 4   No Aerial Apparatus Used at Scene 0 0 2   No Other Vehicles Used at Scene 0 0 9

H 2 No Incident related Injuries   Fire Service 0 0 0   Others 0 0 7   No Incident-related Fatalities   Fire Service 0 0 0   Others 0 0 0   Complex HOTEL 1 4 4

I Fixed Property Use HOTEL   4 1 5   Mobile Property Type NOT APPLICABLE   1 9 8

J Area of Fire Origin BEDROOM   2 1   Level of Fire Orig. Above level 8 9 feet   1 1   Termination Stage in or after flame stage   3

K Equipment Involved in Ignition (if any) NO equipment involved   9 1 8   Form of Heat of Ignition CIGARETTE   3 1

L Type of Material Ignited FINISHED GOODS 1 7 2   Form of Material Ignited MATTRESS   3 1   Ignition Factor FALLING ASLEEP 3 3

M Structure Type 2 or more fixed property use 1 2   Construction Type unprotected ordinary 1 6   Construction Method SITE BUILT   1 1

N Extent of Flame Damage CONFINED TO FLOOR or FOR 4 1 5   Extent of Smoke Damage CONFINED TO STRUCTURE 1 6   Extent of Water Damage CONFINED TO STRUCTURE 1 6

O Extent of Fire Control Damage CONFINED TO STRUCTURE 1 6   Detector Performance OPERATIONAL   1 1   Sprinkler Performance NO SPRINKLERS   1 8

P IF FLAME SPREAD BEYOND ROOM OF ORIGIN   Type of Material Generating Most Flame FINISHED GOODS   1 7 2   Avenue of Flame Travel OPENED DOOR   1 4

Q IF SMOKE SPREAD BEYOND ROOM OF ORIGIN   Type of Material Generating Most Smoke FINISHED GOODS   1 7 2   Avenue of Smoke Travel AIR CONDITIONING DUCTS   1 1

R Method of Extinguishment PRE CONNECTED hose lines with WATER from hydrants.   1 6

S Estimated Total Dollar Loss 0 0 0 0 5 0 0 0 0   Property Damage Classification $50,000 TO $249,999   1 6   Time from Alarm to Agent Application 2 minutes or less   1 2

T Officer in Charge (Name, Position, Assignment) George Ausborn   Date 2-26-87

Member Making Report (if Different from Above) Capt. Roberto Vela   Date 2-26-87

*List name, age, sex, and description of Injury for each casualty on form 902G

**Complete Below

☐ Check box if remarks are made on reverse side

U 3 If Mobile Property | Year | Make | Model | Serial No. | License No. (if any)

V 4 If Equipment Involved in Ignition | Year | Make | Model | Serial No. | Voltage (if any)

COMPLETE ON ALL INCIDENTS

COMPLETE IF CASUALTY OR FIRE

ALL IGNITIONS

COMPLETE IF FIRE FOR STRUCTURE FIRE ONLY

ALL FIRES

COMPLETE ON ALL INCIDENTS

Back of Incident Report

Type of Incident: Structure fire
Location Received: La Posada 100 N. Main
Dispatcher: Juan Robledo

Who Investigated Incident: Fire Marshal Castro
First Unit on Scene: Engine 1
Structure Dimensions: <u>250</u> W <u>250</u> H <u>35</u> H

Weather: Approx Temp <u>55</u> Wind <u>NW5</u> MPH Humidity 100%
Insured by:
Gallons Used: 1600 Light Water Used: 0

Total Man Hours (All Units)   H <u>110</u> M <u>36</u> S <u>0</u>
10-97 (1st Unit):   0414
Control Time: 0619
10-98 (Last Unit) :   1110
10-14 (Last Unit) :   1116

I:
V: 2,500,000
L: 350,000
S: 2,150,000

Remarks:

   On Engine 1 arrival there was smoke and fire visible on the north side
of the building. There were people on the 2nd and 3rd windows calling for
help. Only 1 man was assigned to confine the fire and managed to promptly
knock down most of the fire. The rest of the manpower were assigned to
rescue and evacuate all people from the hotel. A second and third alarm
were called and all were given orders to rescue and evacuate all people on
all floors. This was all done first.

   Mutual aid was called for from Edinburg, Pharr, and Mission. Edinburg
responded with 3 units and 8 men. We used the men to assist evacuation,
and rescue unit lights.

   Pharr responded with 2 units and 12 men. We used Pharr's light and
air and their manpower to overhaul the fire.

   Mission responded with 2 units and 15 men. We used their rescue truck
for air and lights and their men for overhaul and salvage.

nd. Page

0413 hrs.

FIRE DEPARTMENT

DISPATCHER'S INCIDENT REPORT

0414 hrs.

TE:	DAY OF WEEK:	HOW REC'D:	LOCATION REC'D:
-25-87	Wednesday	Code 1	La Posada, room #223

RRECT ADDRESS & ROOM OR APT. NO.		PERSON REPORTING ALARM & PHONE NO:	
00 North Main		Armando Palma	686-6411

YPE OF ALARM: ST (R) (6)(3) 4 5 ALERT! OTHER	**3 ALARM FIRE**	CITY LIMITS (YES) NO	CONTROL TIME: 0619 hrs.

FFICER IN CHARGE:	PERSON SUPPLYING INFO:	INVESTIGATOR:	FIRST UNIT ON SCENE:	SHIFT:
111		200, 221, 220, 222	Rescue 1	C

PORT NUMBER: 0257	CENSUS TRACT: 211	Remarks: Cont'd information from first page.

Hotel Fire.

736 hrs. - Alpha 1 requested Cadets at scene.

710 hrs. - Bravo 3 requested ambulance eastside parking lot, Code 3, Catalina ambulance advised.

DISPATCHER SIGNATURE

Frank M...

TIME OF ALARM:

st. Page

0413 hrs.

CITY OF McALLEN
FIRE DEPARTMENT

DISPATCHER'S INCIDENT REPORT

TIME OF ARRIVAL 1ST UNIT:

0414 hrs.

TE:	DAY OF WEEK:	HOW REC'D:	LOCATION REC'D:
-25-87	Wednesday	Code 1	La Posada, room #223

RRECT ADDRESS & ROOM OR APT. NO.		PERSON REPORTING ALARM & PHONE NO:	
00 North Main		Armando Palma	686-6411

E OF ALARM: ST (R)(2)(3) 4 5 ALERT! OTHER	**3 ALARM FIRE**	CITY LIMITS (YES) NO	CONTROL TIME: 0619 hrs.

FICER IN CHARGE:	PERSON SUPPLYING INFO:	INVESTIGATOR: #'s 200, 221, 220, 222.	FIRST UNIT ON SCENE:	SHIFT:
#111			Rescue 1:	C

PORT NUMBER: 0257	CENSUS TRACT: 211	Remarks:

Hotel Fire.

17 hrs. - 2nd alarm, advised water plant.

22 hrs. - 3rd alarm, mutual aid requested from Pharr, Edinburg, & Mission.

17 hrs. - Charlie 4 requested ambulance, Main St. Side of Parking Lot, Catalina ambulance advised.

46 hrs. - Bravo 5 requested ambulance.

(cont'd)

DISPATCHER SIGNATURE

Frank M...

UNITS RESPONDING	NO. OF PERSONNEL	TIME OUT	TIME CHECK	IN SERVICE	ASSIGNMENT COMPLETED	IN STATION	MAN HOURS H	M	MILES
Bra. 3	1	0430 hrs.	None	None	None	None	N/A	N/A	N/A
Bra. 4	1	0430 hrs.	None	None	None	None	N/A	N/A	N/A
Miss.	N/A	0422 hrs.	0447 hrs.	0830 hrs.	---------	--------	N/A	N/A	N/A
Pharr	N/A	0442 hrs.	0449 hrs.	0700 hrs.	---------	--------	N/A	N/A	N/A
Edbg.	N/A	0442 hrs.	0445 hrs.	---------	---------	--------	N/A	N/A	N/A
Bra. 5	1	0430 hrs.	0504 hrs.	None	None	None	N/A	N/A	N/A
Eng. 2	3	0417 hrs.	None	0643 hrs.	None	0852 hrs.	7	18	5

OFF DUTY									
TOTAL	36						110	36	28

NUMBER: 0257   **CORRECT ADDRESS:** 100 North Main

UNITS RESPONDING	NO. OF PERSONNEL	TIME OUT	TIME CHECK	IN SERVICE	ASSIGNMENT COMPLETED	IN STATION	MAN HOURS H	M	MILES
Res. 1	2	0413 hrs.	0414 hrs.	1044 hrs.	1110 hrs.	1116 hrs.	13	54	1
Eng. 1	3	0413 hrs.	0414 hrs.	0900 hrs.	None	0903 hrs.	14	21	1
Eng. 6	3	0413 hrs.	0414 hrs.	1018 hrs.	1043 hrs.	1048 hrs.	19	30	1
Cha. 4	1	0413 hrs.	0414 hrs.	0932 hrs.	None	0936 hrs.	5	19	1
Eng. 3	3	0417 hrs.	0420 hrs.	0842 hrs.	None	0850 hrs.	13	15	4
Eng. 4	3	0417 hrs.	0423 hrs.	0900 hrs.	None	0905 hrs.	14	09	4
Lad. 4	2	0413 hrs.	0417 hrs.	0537 hrs'.	None	0937 hrs.	2	48	4
Eng. 5	3	0417 hrs.	0422 hrs.	0535 hrs.	0842 hrs.	0853 hrs.	13	15	7
Alp. 3	1	0417 hrs.	0431 hrs.	None	---------	---------	N/A	N/A	N/A
Lad. 1(Y-1)	1	0433 hrs.	0436 hrs.	0721 hrs.	---------	0735 hrs.	N/A	N/A	N/A
Alp. 1	1	0417 hrs.	None	None	None	None	N/A	N/A	N/A
Bra. 1	1	0430 hrs.	0510 hrs.	1117 hrs.	---------	---------	6	47	N/A
Bra. 2	1	0430 hrs.	None	None	None	None	N/A	N/A	N/A
OFF DUTY									
TOTAL	Next Page						Next	Page	N/P

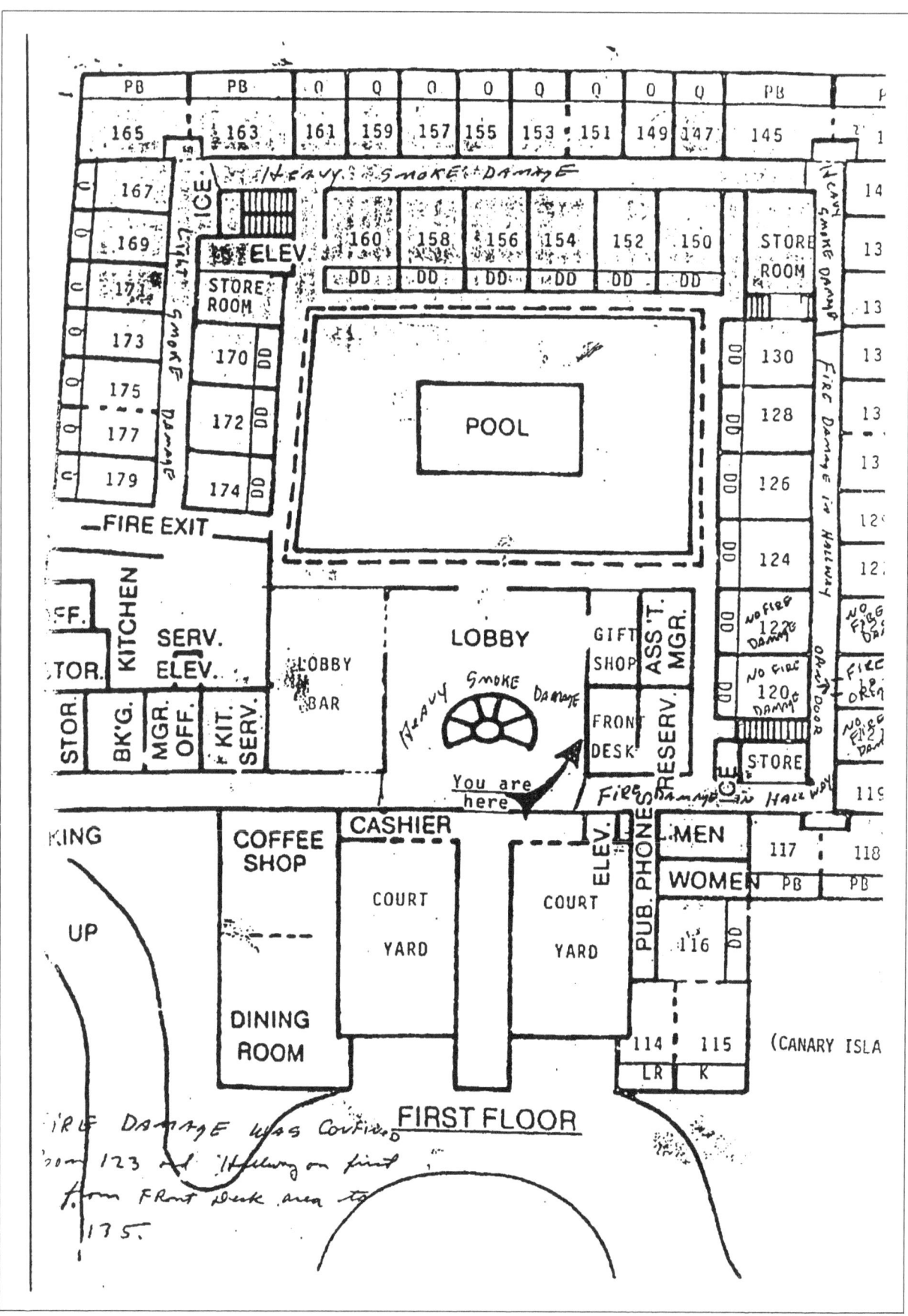

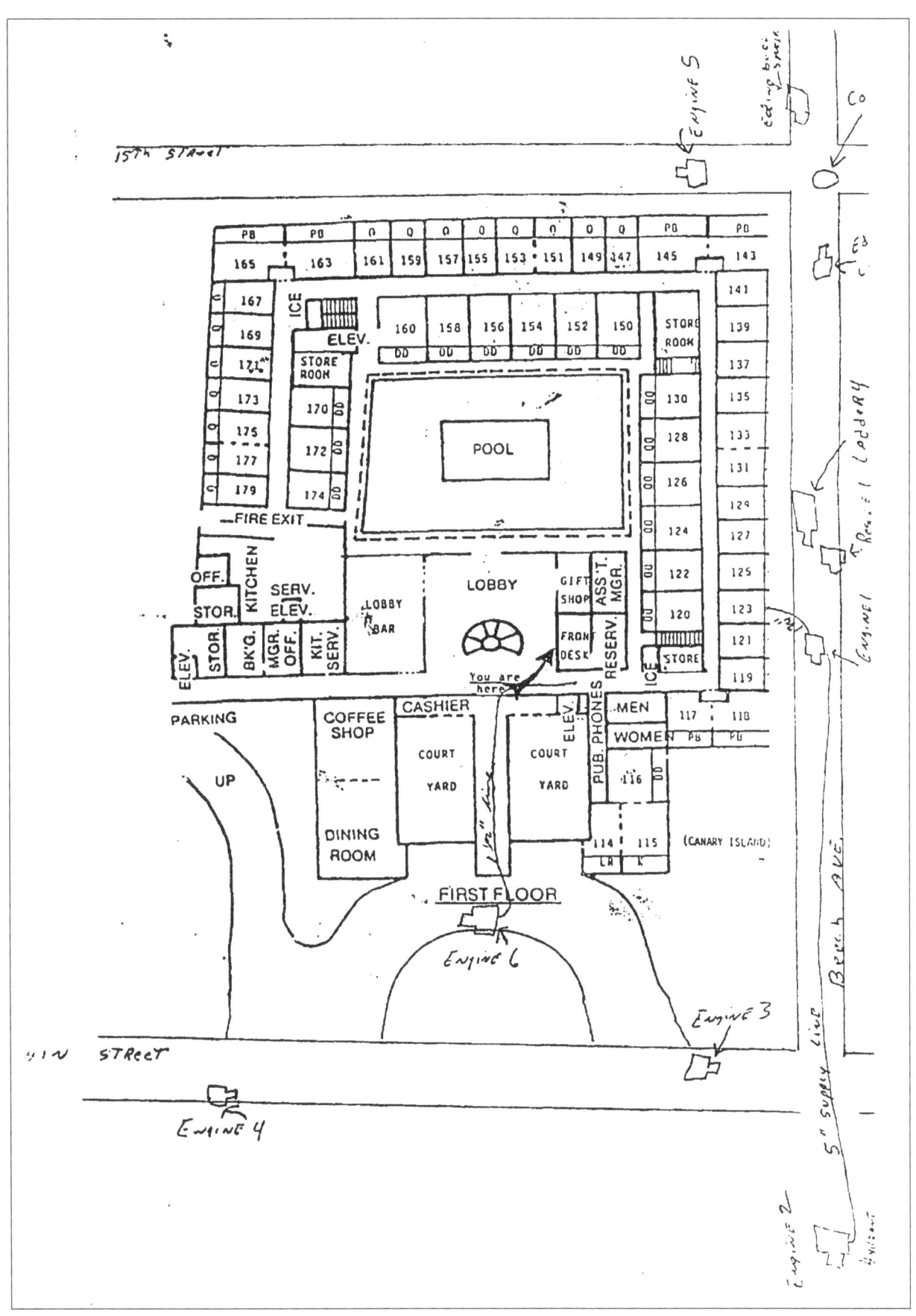

TO: Deputy Chief, George Ausborn

FROM: Lieutenant Juventino Villarreal

DATE: February 26, 1987

SUBJECT: La Posada Hotel Fire

On February 25, at 0413 hours, an alarm was dispatched at La Posada Hotel room 223. Engine 6 arrived at the north side and proceeded to the east side and parked in front of the entrance. There were people on the balconies of the second (2nd) and third (3rd) floors. We advised them to stay calm and proceeded to get them down. There were approximately (7) occupants.

Then pulled the 300' line and advanced the line to the main entrance. The line was left charged for the incoming units to do the firefighting.

We then proceeded to the north side of the third floor where the concentration of smoke was heavier and began the evacuating of occupants room by room. We evacuated approximately (6) occupants. Engine 6 with the help from Engine 3 evacuated all the third floor then I advised Captain Vela that the third floor was clear.

Then we proceeded with the north side of the 2nd floor and at this time, one firefighter and I were exhausted from knocking doors open so I sent him to get a master key. Approximately a minute later he arrived and the evacuation continued. At this time, we were joined by Engine 5 crew. On room 235, we knocked the door open and found a couple sleeping and advised them to evacuate because of the heavy smoke and they said they were going to get some clothes on. And so I left a firefighter there to escort them out. Then we found a group of about (12) occupants in the inside balcony and we escorted them out of the building. Then, we continued but found that the south part of the second (2nd) floor had been evacuated so I made a quick search of the south side and then advised Captain Vela that the 2nd floor was clear. Then Engine 6 crew with the people working overtime made quick and thorough search of the third (3rd) floor again and posted a guard at the south exit leading to the parking lot.

On my second search of the second floor I found out that the people on room 235 did not want to evacuate so, I went and got two Police Officers and they evacuated these people.

Then I continued and posted a guard on the second floor south exit and proceeded to the lobby area. I was advised that the fire was out and to go take a break.

After the break we helped Chief Garza with the process of taking occupants to their rooms to get their belongings, then we were relieved.

Sincerely,

Lt. Juventino Villarreal

TO: Deputy Chief, George Ausborn

FROM: Lieutenant Juan G. Menchaca

DATE: February 26, 1987

SUBJECT: La Posada Hotel Fire

Sir:

The following is an account of Engine 5 and crew's role in the 3rd alarm incident at La Posada Hotel on February 25, 1987:

With a one minute estimated time of arrival, I radioed Whiskey 1 and requested assignment for Engine 5. Whiskey 1 assigned Engine 5 to the west side of the complex's exterior to commence evacuation of the 3rd floor. A secondary assignment was to enter and perform search and rescue. Upon our arrival we set up the Tele Squirt's boom and proceeded to perform rescue' evacuation with the aerial ladder on Engine 5 in company with Ladder 4. Once the situation on the 3rd floor's west side was under control, Firefighter del Angel and myself entered through room 345's balcony window. We then commenced search and rescue of the 3rd floor. The west hallway rooms were searched with several requiring forcible entry. The same operation was carried out on the north hallway rooms. Several rooms were found to be occupied and required. to escort the occupants out to safe areas. The operation to search and rescue was carried out until satisfaction came that the 3rd floor was secure. Once control of the situation on the 3rd floor was accomplished, Engine 5's crew worked on getting our unit back in service. Once Engine 5 was placed 10-8 at the scene, our crew was assigned to Charlie 2 to assist with the hotel occupants. This assignment required that we escort the room's occupants through a check-point and up to their rooms. Once at the room, the occupants were allowed to gather and remove their personal belongings. After this we were ordered to then escort them out through the check-point. This assignment was carried out until released to return to Fire Station 5.

Sincerely,

Lt. Juan G. Menchaca

Lieutenant Juan G. Menchaca

TO: Deputy Chief, George Ausborn

FROM: Lt. Guadalupe Castillo Jr.

DATE: February 26, 1987

SUBJECT: Actions Taken at La Posada Hotel

     Upon receiving the alarm, Engine 4 crew consisteing of (3) men responded
to the fire. When arriving at the scene we entered the main lobby. There was
some fire above the front deck and the hallway at the Northeast corner of the
hotel. A hose line was in place but little or no water pressure at the time.
Capt. Vela called on the radio saying to start Rescue Operations. Receiving
the orders we started the Rescue Operation at once. We took the second
floor and started on the north half working our way west around the hotel. We
found some persons and took them out of the hotel. All the rooms were checked
by opening the door and after checking the room a chair was placed in the
doorway. The number of persons taken out of the hotel were over (10) and
still more persons were taken out from the south side of the. second floor.
     All of my actions were Rescue Operations and no fire fighting on the
second floor.
     I did notice that the fire was under control when we were taking persons
out through the main lobby.
     After a control time was given I helped the hotel guest go back to their
rooms and watch them pack their clothes and leave the hotel.
     We the crew of Engine 4 returned to our station and were off duty for the
day.

                              Sincerely,

                              H. Guadalupe Castillo Jr.
                              Lt. Guadalupe Castillo, Jr.

TO: Deputy Chief, George Ausborn

FROM: Capt. Roberto Vela

DATE: February 26, 1987

SUBJECT: La Posada Hotel Fire

After receiving an alarm for Room 223 on fire at La Posada Hotel, we responded to east on Highway 83. As we turned north on 15th Street, I could see some smoke rising from the roof. I told my driver to go around and turn on Beech so that I could size up the situation, as we turned on Beech, We could see some people on the upper windows calling for help. We also saw fire rolling out of a window on the first floor. Engine 6 and Rescue 1 were right behind me. I ordered Engine 6 to go in on the front entrance and start rescue operations from the inside. I ordered my firefighter to contain the fire with a 1 314" line, then go in through the window of the room that was totally involved. Rescue 1 crew and Engine 1 crew that were left, started rescue problem. I then called for a second alarm. The second alarm consisted of Engine 3 and Engine 2. I ordered Engine 3 to go in the front entrance and start rescue operations on the third (3rd) floor. I ordered Engine 2 crew to go in the front entrance and start evacuating the building. I called for a third alarm and ordered Engine 5 to set up their ladders on the west side of the building so they could start bringing people down. I ordered Engine 4 crew to go to the third floor and assist with the rescue operation.

After I had assigned all the companies, I advised Charlie 4 that I was going into the building to check out the situation. I went into the building to check out the situation. I went into the building and found that everything was going very smoothly. The fire had been knocked down on the hallway and all the floors were being evacuated.

Sincerely,

Capt. Roberto Vela

Capt. Roberto Vela

TO: Chief, George Ausborn

FROM: Hector Cantu

DATE: February 26, 1987

SUBJECT: La Posada Hotel Fire

When we arrived, Capt. Vela advised me to drive to the north side of the building. When I stopped the truck in front of the room that was on fire, Capt. Vela advised Firefighter Castaneda to use the 1 3/4 hose. I looked up to the building and saw several people in the windows yelling for help. I advised the people not to panic or jump. I also told them that the fire was out and that there wasn't any danger to them. That's when I saw Miss Johnson fall/jump out the window onto the ground. Miss Johnson was sitting on the edge of the window so I can't say if she jumped or fell out the window. After she fell/jumped, Rescue 1 personnel and I put her on a back board and into the ambulance. We then put the 24' extension ladder to assist another lady down off the 3rd floor, but it didn't reach. We then went to Ladder 4's ladders and got the 32' extension ladder. Before we put the 32' extension I went up the 24' extension and advised the lady not to panic or jump and that the fire was out. We then extended the 32' extension ladder up to her window. I then went up and got the lady down.

Sincerely,

Driver,
Hector Cantu

TO: Chief, George Ausborn

FROM: Juan Reyes, Firefighter, A/Driver on Rescue I

DATE: February 26, 1987

SUBJECT : La Posada Hotel Fire

On the morning of February 25, 1987, I Juan Reyes and my partner Ruben Ramos, responded to a regular alarm at La Posada Hotel for a room fire. While enroute to the scene, we observed smoke coming from the building. Upon arriving at the north side of the building, I saw a person on the 3rd floor that was yelling for help. I yelled to her to stay calm and that the fire was being controlled. While I was telling the lady to stay calm, my partner went to Engine 1 to get an extension ladder to help people from the third (3rd) floor get down. At this time I saw the lady falling from the window where she had been. As soon as she hit the ground, I immediately went to her to render aid. She was conscious and telling me to move her out of the way. She kept telling me "help me, aren't you going to help me?". I advised her to stay calm and not to move so as not to injure herself any further. I stayed with her until my partner arrived and I went to get a back board. Upon arriving with the back board, myself, my partner, Ruben Ramos, and Driver Hector Cantu, loaded the lady on the back board and immediately into the ambulance which had just arrived. The three (3) of us continued to help people down from the third (3rd) floor windows.

Sincerely,

Juan Reyes

Firefighter, A/Driver on Rescue I
Juan Reyes

TO: Chief, George Ausborn

FROM: Lt. Gilbert Longoria

DATE: February 26, 1987

SUBJECT: La Posada Hotel Fire

    When giving 2-alarm for fire on hotel, we responded down Main St. all
the way to fire scene. Captain Vela advised us to don air paks and go inside
hotel for search and rescue. On the corner of Main and Beech firefighter
(Davila) and myself, dropped out to go to fire scene. I advised driver
(Salinas) to supply water line to Engine 1, which was parked between 14th and
15th Street on Beech St. Then, firefighter and myself went in through the
main entrance of hotel, advised firefighter to go to second floor to start
search and rescue. I grabbed 14 line from Engine 6 to extinguish fire on
front desk and the hallway. Fire was put out and I started to search and
rescue on the first floor. I searched from 125 to 179, as to extinguish fire
which was igniting again on different spots. Firefighter (Davila) went to second
floor through stairway and met with Engine 3 crew and started to search and
rescue that floor. They found some people still in rooms and were rescued down
to the first floor. Firefighter and Engines 3, 5, and 6 went to the 3rd floor
to search and rescue. And firefighter said they found a lot of people in rooms
an the people didn't know there was a hotel fire. We had to knock down doors
to make sure no one was inside, cause some people didn't want to open doors
After a while, I had manpower which started to knock down ceiling on front
desk and all the hall way from room 117 to 135. And then, Chief Garza called all
crews to lobby and told us we were going to help the people with their belongings.

                          Sincerely,

                          Lt. Gilbert Longoria

TO: Deputy Chief, George Ausborn

FROM: Lieutenant Juan Palacios

DATE: February 26, 1987

SUBJECT: La Posada Hotel Fire

Engine 3 crew "C" Shift
2-25-87   @   0417 hrs.

    Call came in for a room on fire at La Posada - Rm. 123. I told my men to get ready because it was a great possibility of being a second alarm fire. When second alarm was activated Engine 3 crew was already leaving station. While enroute Captain Vela gave me orders to evacuate third (3rd) floor. On arrival, we parked on east front entrance and I told my driver to stay with the truck and my firefighter and I proceeded to enter building on east side main entrance. As we entered we saw flames to our right side on first floor by front desk and also down north hallway. I told my firefighter to forget about fire and to go to the third (3rd) floor to evacuate people. My firefighter and I started banging and yelling on doors; no one would open doors so we started kicking in doors and found people inside rooms. We took them out the main spiral stairs and out the front (east side) of the building and Driver Palacios and Gutierrez were assisting people to safety. These are the rooms that Engine 3 crew checked: Room 102, Tower, Rooms 355 through 392, Ball Room, Valencia Room, Storage Room, Bar, panting all these on third (3rd) floor. When we went outside to change air bottles, Driver Gutierrez advised me that a lady was trapped in a room on the first floor, room 117. My firefighter went to outside window to remove burglar bars and I went inside to break in. I took lady out through front entrance. Engine 3 supplied 1000 gallons of water to Engine 6. We evacuated approximately 50 people from third (3rd) floor. By this time Edinburg crew was sent up stairs to assist me and I ordered them to recheck all rooms again and to set markers on all doors on third (3rd) floor. Engine 3's crew last assignment was to assist people in checking out from rooms.

                                          Sincerely,

                                          Lieutenant Juan Palacios

FIRE CHIEF
E. H. DERR

FIRE MARSHALL
HONORE CASTRO

DEPUTY CHIEFS
JOE GARZA
ODILIO MONTES
RENE DEL BOSQUE, S
GEORGE AUSBORN

CITY OF McALLEN
FIRE DEPARTMENT
(512)631-3301   101 S. BICENTENNIAL   McALLEN, TEXAS 785001

March 2, 1987

Mr. Leo B. Womack
Womack/Gilman Interests
1001 Texas, Suite 600
Houston, Texas 77002

Dear Mr. Womack:

Under The Life Safety Code 101, 1981, the following requirements will be requested for The La Posada Hotel, 100 North Main, McAllen, Texas.

1. Sprinkler system in basement and high hazard areas.
2. Complete fire alarm and smoke detection system. (State Certified).
3. Stand pipe system throughout with hose cabinet.
4. Enclose stairwells with fire resistant materials.
5. Secondary exits from basement and towers.
6. Enclose boiler room area with two hr. rated walls.
7. Automatic shut of in air handlers.
8. Emergency lights throughout.
9. (5) lb. A.B.C. Fire Extinguishers.
10. Correct any and all penetrations.
11. Automatic door closures for all doors.
12. Correct existing emergency lights.
13. All electrical equipment must be done in conduit.
14. Remove decorative/burglar bars.

The following are recommendations:

1. Sprinkler system throughout.
2. Intercom.and P.A. system.
3. Evacuation plan and emergency procedure in each room.
4. Emergency plan.
5. Fire drills.

TO: MR. LEO B. WOMACK
DATE: March 2, 1987
SUBJECT: LA POSADA HOTEL (REQUIREMENTS AND RECOMMENDATIONS)
PAGE 2

It is understood that some of the requirements and recommendations may need an explanation. Please feel free to call or come by our office.

Keep in mind that there may be other requirements and recommendations to follow. Thank you for your cooperation.

Sincerely,

Honore Castro
Fire Marshal
Fire Prevention Division

HC/bl

cc: B.J. ULCAK, LA POSADA GENERAL MANAGER
      RICHARD HINOJOSA, ASSISTANT CITY MANAGER
      FERNANDO ROMEROS, DIRECTOR OF BUILDING INSPECTIONS